WITHDRAWN

The Spheres of Life

Also by JOSEPH W. MEEKER

The Comedy of Survival:
Studies in Literary Ecology

The Spheres of Life
An Introduction to World Ecology

JOSEPH W. MEEKER

Illustrated with drawings

Charles Scribner's Sons · New York

Copyright © 1973 Athabasca University
Copyright © 1975 Joseph Meeker

The quotation on page 63 is from "Puzzle" by Kenneth E. Boulding, from *Future Environments of North America*, edited by F. Fraser Darling and John P. Milton, copyright © 1966 by The Conservation Foundation. Used by permission of Doubleday & Company, Inc.

Library of Congress Cataloging in Publication Data
Meeker, Joseph W
  The spheres of life.

  Bibliography: p.
  1. Ecology. I. Title.
QH541.M388   574.5   74–11217
ISBN 0–684–13937–5

This book published simultaneously in the
United States of America and in Canada—
Copyright under the Berne Convention

All rights reserved. No part of this book
may be reproduced in any form without the
permission of Charles Scribner's Sons.

1 3 5 7 9 11 13 15 17 19 v/c 20 18 16 14 12 10 8 6 4 2
Printed in the United States of America

For my mother and father

Acknowledgments

This book was originally prepared to supplement an interdisciplinary course in World Ecology offered by Athabasca University, Edmonton, Canada. Its chapters first appeared as a series of articles published by *The Edmonton Journal* in the fall of 1973.

Scientific advice was provided by an interdisciplinary team of Athabasca University faculty, but especially by Dr. T. S. Bakshi and Dr. Rae Laurenson. Carolla Christie and James Roebuck contributed editorial assistance. Joyce Clark prepared the glossary with the help of my wife Marlene. Brenda Egglestone proofread the final copy and compiled the index.

Rick Pape and Sally Don drew the illustrations; the circular symbols used as chapter headings were created by Greg Prygrocki.

Contents

Introduction	3
1. Systems of Life	7
2. The Lithosphere	15
3. The Hydrosphere	23
4. The Atmosphere	31
5. The Biosphere: Plant Life	39
6. The Biosphere: Animal Life	47
7. The Noosphere	55
8. Paths of Energy	63
9. Populations	71
10. Evolution	79
11. The Principle of Diversity	87
12. Environmental Ethics	95
Glossary	105
Bibliography	111
Index	117

Illustrations

Interrelated ecological systems	9
Stages of soil destruction	17
The hydrologic cycle	25
Pollutants in the atmosphere	35
Successive stages of plant life	43
Wolf clan—normal animal relationship	49
Man's self-created maze	59
The candle of civilization	65
Effect of population pressure	73
Five hundred million years of evolution	81
Value of diversity	89
Perils of manipulation	97

The Spheres of Life

Introduction

During a million or more years of the development of the human species, the ecological environment was the basis for everyone's education. Hunting and agricultural peoples have always known that the study of the land, plants, animals, and their intricate relationships must provide for every young person the knowledge essential to build a good life. Before there were schools and universities, the natural environment offered experience and insight sufficient to build a meaningful and well-oriented life.

In recent centuries educators have been concerned only with the study of human affairs, past and present, as if a knowledge of nature were unnecessary. When nature has been studied, it has

THE SPHERES OF LIFE

often been presented as a "resource" to be used for human purposes. Schools have become places where people are taught how to control and to exploit nature, or are trained in science as a discipline of the human mind, but rarely are students taught how to participate knowingly in the ecology of which they are a part. Young members of our species have thus learned an unprecedented style of living based upon the pretense that human beings can live by rules of their own without regard for the natural order of things. Only a few generations of such living habits have been enough to create the environmental crisis which now threatens the integrity of nature and of one of nature's more interesting experiments, mankind.

It is time, then, to begin all over again in the hope of rediscovering the ecological environment as a basis for education. However, we cannot be content, as our ancient ancestors could, to acquaint ourselves merely with the local environment in which our personal lives are to be lived. Humans are now a global species with global awareness and responsibility. Humanity is part of the world ecosphere.

This book offers an introduction to the ecology of the world and a glimpse of some of the roles available to mankind within it. The origins of some ecological problems that are attributable to human thoughts and actions will be identified, and some prospects for more harmonious integration of hu-

Introduction

manity with natural ecology will be suggested. The purpose of the book is to arouse curiosity about the subjects discussed in the hope that this will lead to further study of the human role in world ecology. The future of our own and other species depends upon humanity's ability to live compatibly with the natural world.

1. Systems of Life

Human civilization managed to get along for thousands of years without professional ecologists. Why is this new science suddenly on everyone's lips and in headlines everywhere? Perhaps it is because the world has begun to experience an environmental crisis of a sort that has never before occurred. Just as the science of psychology was created early in the twentieth century to respond to widespread problems of mental disease, so ecology has emerged in our time because the earth and its systems are deeply disturbed and require a new kind of attention.

Like other diseases, ecological crisis must be understood and treated if we hope for the health of humanity and of the other forms of life with which

THE SPHERES OF LIFE

humans share the earth. Those who feel hope and responsibility for a better environment need to understand the ingredients necessary for ecological health in order to work intelligently toward its restoration.

The only known setting that will permit the existence of life is found on the planet earth. Unique conditions necessary to living organisms exist here because of the peculiar way in which the earth's various systems are related to one another. There are five such systems which must be in relative harmony if life is to prosper on earth: land, water, air, animal and plant life, and human culture. These systems will be described in later chapters but are summarized briefly here.

The lithosphere is the earth itself with its surface layers of soil and rock and its partly unknown depths, which now and then remind us of their presence by ruffling the surface with a volcano or an earthquake. Most humans have confined their interest in the lithosphere to the skin of its outer crust, for that is where they walk, build their homes and highways, and hunt or farm for their food.

The earth's surface is the basic setting for life of all kinds. Its varied soil chemistries dictate the kinds and numbers of plants that will grow. Its mountains influence weather patterns and the distribution of moisture through rainfall and runoff. Every living thing is limited by the nonliving land

The ecology of the world is a complex process of interrelated systems, each overlapping and integrated with others. Sky, land, and water touch and intermingle in ways that support plant and animal life, and together these constitute the interlocking world ecosphere.

THE SPHERES OF LIFE

forms which constitute the terrestrial environment. And every living thing makes some contribution, creatively or destructively, to the chemistry of the land upon which it lives.

In the past century mankind has been increasingly interested in probing beneath the earth's surface for access to what is stored there. Extensive mining for mineral deposits and deep wells for water and oil have extended man's influence to levels of the lithosphere that have never before been affected by human activities. So far, the full consequences of these new penetrations into the earth are largely unknown and unpredictable. Of all the component systems of life, probably least is known about the earth itself.

The hydrosphere is the waterworks of the world and the most unusual physical feature of this planet. It is the system which stores, distributes, and purifies water, that miraculous compound without which no plant or animal can exist. Since water can assume the qualities of a gas, a liquid, or a solid, it can move readily from the atmosphere to the lithosphere and back again, pausing now and then in the form of ice to cool a summer beverage, to block a winter street with snow, or to form part of a polar icecap. Glaciers made of frozen water gouge deep valleys in the land, and rivers are capable of dismantling whole mountains and moving them a piece at a time to valleys and to the sea. In its quiet way,

Systems of Life

the hydrosphere rearranges the earth from day to day, although it is also capable of noisier and more dramatic displays such as floods and violent waves.

Since water moves readily among the other systems of the earth, it is the ideal medium for transporting portions of one system into another. Wastes from living processes find their way into the earth, and chemicals from the earth enter living matter thanks to water. Whatever is consigned to water achieves a new form in some new place.

The atmosphere is the blanket of gases that wraps the earth, including the air needed by plants and animals and the many other vapors and gases which accumulate in the sky. It is both a storehouse and a distribution system for the oxygen produced by green plants. Its winds transport water and bring weather, fill sails, power windmills, and ruffle the hair of fastidious ladies.

Like water, the atmosphere is capable of tantrums which release enormous destructive energy. It can also be a purveyor of poisons such as toxic chemicals or radioactive particles, for it does not care what it carries and it accumulates whatever is added to it, just as cigar smoke collects in the air of a closed room. This would be a stuffy world indeed if the atmosphere were not large enough to disperse most of its collected chemicals.

The atmosphere is the roof of the earth, shielding its surface from solar and cosmic radia-

THE SPHERES OF LIFE

tion, from meteors which are usually consumed by atmospheric friction, and from the extreme temperatures which afflict planets not equipped with atmospheric insulation.

The biosphere is organic life in all its forms, from a single cell to the most complicated of plants and animals. The intricate relationships among the various forms of life as they interact with the non-living systems previously described are studied by the science of ecology. Food chains and webs, the structures and functions of living systems, the evolutionary development of plant and animal species, and the delicate conditions necessary to the perpetuation of organic life constitute the ecological study of the biosphere. No single element exists in isolation; here, more than elsewhere in nature, everything depends upon everything else.

The biosphere has produced human life and has sustained its development over many thousands of years. Human bodies are animal bodies, depending like other animals upon water, air, food, and upon the continuous existence of many other forms of plant and animal life. To the extent that the earth's biosphere is diseased or endangered, human life is jeopardized as well.

The noosphere (from the Greek word *Noos*, the mind) is that part of human life that is not shared in common with other animals. It is the human mentality and its collective product, cultural

Systems of Life

civilization. The noosphere is the domain of distinctively human mental functions such as abstract thinking, the creation of symbols, and the powers of the human imagination. It includes cultural and spiritual traditions, art and science, philosophy and religion, technology, and the human capacity to learn and to record knowledge as no other species is able to do. The noosphere is the intricate system of human beliefs and values painstakingly developed over the past 6,000 years of recorded cultural history.

The environmental crisis of the twentieth century is largely a consequence of the noosphere's influence upon the biosphere and upon the other great natural systems of the earth. Humans have used the earth as if life did not depend upon it, but it does. Now we must look within the human past and within the modern human mind for new forms of knowledge and belief to bring the human noosphere into closer compatibility with the other systems essential to life on earth. Public awareness of ecological problems can be the first step in this direction, but many steps remain to be taken.

These five great spheres together constitute the world ecosphere. If any one of the ecosphere's systems is significantly incompatible with the other four, serious and sometimes destructive consequences are sure to follow. If political factors in the noosphere dictate that a nuclear explosion must oc-

THE SPHERES OF LIFE

cur in Tahiti, the effects will be evident in the air, water, and land of Japan, and in the milk delivered to North American doorsteps. Environmental crisis is the collective result of many such events which cause imbalance in the world ecosphere.

2.
The Lithosphere

    Humans like to speak of the earth as if it were their property. We think of it as "our" earth, and we judge its various environments as good or bad according to their ability to satisfy human needs and desires. Given a human brain, rattlesnakes might make similar judgments, though their definition of land values would differ greatly from human preferences. The truth is that the earth is no one's property. It is a complex system of physics and chemistry showing little regard for the kinds of life which have appeared in the past few million years on its outer skin. The earth does not need life, but life needs a whole earth. Humans are the property of the earth.
    Soil is created when rocks are exposed on the earth's surface and begin to decompose into small

THE SPHERES OF LIFE

particles because of the action of air, water, and sunlight. Like rust upon a mountain of iron, the soil appears as a tiny and superficial feature on the great mass of the earth. Decomposed rock is transformed into fertile soil by the addition of plant and animal wastes, and soon supports a complex microscopic flora and fauna of fungal and bacterial life. Rich soils are a cooperative product of all the spheres of life acting in concert.

The location and chemistry of soils are of small importance to the lithosphere but matter very much to the creatures living on the earth's surface. Trees and grasses bind the soil and hold it in place, and add to it the enriching organic products of their life and death. Once established and stabilized in this manner, the soil is ready to receive a wide range of plants. Human agriculture is a technique for using the relationship between plants and the soil for the advantage of one species, mankind. Historically, farming has significantly altered the location and composition of soils.

Clearing trees from hillside slopes is an ancient practice. Wood is always useful for building and burning, and the cleared land upon which it formerly grew then becomes available for farming. Several generations of men can raise crops on former forest lands before the marginal fertility of such soils is lost. Domestic animals can then be pastured there until low-growing vegetation has been grazed

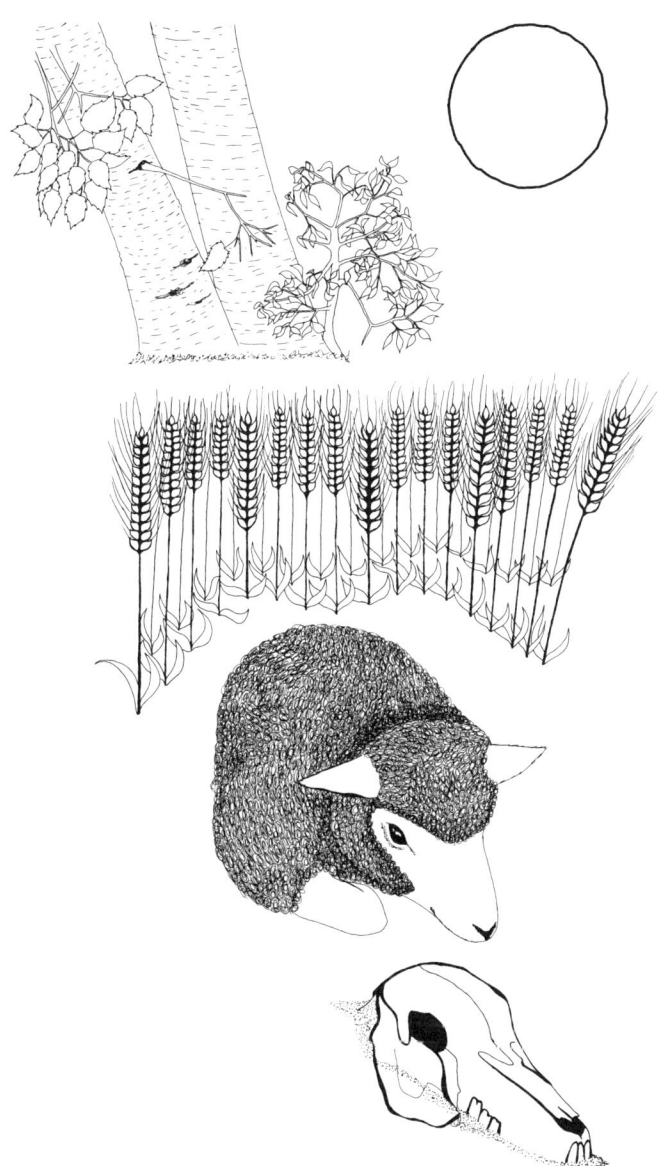

Soil destruction has often proceeded through removal of forest cover, followed by crop farming, then by grazing of cattle and sheep, until all vegetation has disappeared. The remaining poor soil is easily lost to erosion, leaving only desert land with little capacity for sustaining life.

THE SPHERES OF LIFE

away. When even grazing is no longer possible, men and their animals normally move on to repeat the same procedure elsewhere. But the abandoned soil moves on, too, carried by erosion into the valleys.

The encouragement of erosion by removing natural vegetation was mankind's earliest experiment in massive earthmoving, and it has profoundly altered the face of the land. Once the cedars of Lebanon held mountain soils in place just as the cedars of British Columbia do today. But most of Lebanon's mountains are now eroded and bare, their cedars long ago used to build cities. Many of the cities, too, have disappeared beneath the silt that washed from denuded mountainsides.

Some of the great deserts of Asia, Africa, and the Middle East were forests and grasslands before humans began to farm them some 8,000 years ago. The biblical wilderness of Sinai was a genuine wilderness with wildlife and trees until farmers and shepherds in ancient times transformed it into the Sinai Desert. When deep-rooted perennial grasses are replaced by the annual grasses called cereal grains—wheat, barley, rye, corn, rice—two major changes occur in the soil: its chemistry becomes simpler, and its physical stability is decreased because of the shallower roots of domesticated plants. Both changes make it easier for the soil to wash away or blow away, eventually leaving nothing but sterile desert.

The Lithosphere

North American soils have felt the effects of agriculture for only a century or two, yet the processes of soil erosion and depletion have become evident even in that brief time. Forests that required many years to cut in former years are but a day's work for modern mechanized loggers. The floods and dust storms periodically visiting the prairies do the same kind of damage now that they did long ago in Lebanon and other Middle Eastern countries. Experts note that many farming soils in North America are several feet shallower than they were fifty years ago, and the lost soils can be found in the growing deltas of our major rivers. North America is a long way from becoming a desert, but much of its development for the past century has led in that direction.

All the scrapings of mankind, whether with hand tools or bulldozers, can merely rearrange the topmost layer of the earth's crust. Man's most energetic efforts for thousands of years have had but small effect upon the unseen mass of the lithosphere. Only recently have humans been able to penetrate to deeper levels, and already they have created problems there.

Holes several thousands of feet deep have been drilled in recent years to dispose of liquid nuclear wastes and other dangerous substances. Observers in Colorado and Alaska have noted a sharp increase in the number of earthquakes in the vicinity

THE SPHERES OF LIFE

of such holes. Evidently, liquid wastes tend to lubricate the joints between rock layers, permitting them to slip more easily and to extend existing fractures. Some experts believe that such small earthquakes may be desirable because the pressures they gradually release would otherwise accumulate to cause large earthquakes, while other scientists interpret manmade tremors as ominous portents of greater quakes to come. The draining of oil and gas deposits deep within the earth may have similar consequences in some parts of the world. Few experts mention the long-term effect of pumping liquids into or out of deep levels of the lithosphere, for little is known of this subject.

Artificial lakes also increase earthquake activity. The weight of large bodies of water puts stress upon underlying rock layers, and seeping moisture beneath such lakes makes slippage along fault lines more likely. In the past century the human passion for building dams has no doubt caused the rattling of many areas of the earth, in some cases significantly. The famous earthquake of 1906 that destroyed San Francisco occurred only a few years after a large reservoir had been built nearby, and there is good evidence that the two events were connected. Many such hazards may exist near the thousands of manmade lakes created in the past few decades.

The Lithosphere

Careless housekeepers that we are, we have lately acquired the bad habit of sweeping our radioactive dust under the bedrock. Nuclear wastes are often stored in old mine shafts or pumped into concrete pits. Occasionally such pits leak, as one recently did near Hanford, Washington, releasing strontium, cesium, and plutonium into the earth. Deep nuclear explosions are also used to open new gas and oil deposits in buried shales, and atomic testing proceeds underground on the assumption that it will be "safe" there. All of this is done with little knowledge of the influence such potent rubbish may exert a century or a millennium from now upon the hidden chemistry and physics far beneath human feet.

The consequences of mankind's new ability to meddle in the affairs of rocks are largely unpredictable. Little more is known now about the conditions necessary for stability in the lithosphere than ancient man knew about the ecology of the surface. If underground activities are guided by no better wisdom than that of the traditional use of surface lands, the problems of the lithosphere can be expected to multiply rapidly in coming years.

Most of the labor of mankind since the beginning of civilization has been used to modify the surface of the lithosphere. The cumulative result of these efforts has been to convert complicated but

THE SPHERES OF LIFE

stable land forms into relatively simple and unstable forms. This sad history, plus man's recently acquired ability to influence the subsurface systems of the earth, constitutes the major problems of the lithosphere as they affect the conditions necessary to life.

3.
The Hydrosphere

The significance of water cannot be understood by purely scientific investigation, for attitudes about water and its use have been shaped by many non-scientific influences. Philosophers, for instance, have long pondered the apparent contradiction that water's strength arises from its weakness. Water tends always toward harmony with its surroundings, assuming the shape of whatever contains it and changing its state from liquid to vapor or ice according to the prevailing temperature. It seeks the lowest possible places for itself and accepts whatever is done to it or placed in it. Water is thus symbolic of humility before the earth and of adaptation to given environmental conditions. Yet it is also a symbol of power, for water is capable of moving mountains

THE SPHERES OF LIFE

into the sea, splitting rocks, carving valleys, and causing death and destruction. Water can give life and take it away.

The belief that water is able to cleanse itself of harmful substances has made it a traditional symbol of self-purification in many world religions. Under natural conditions, moving water does tend to distribute and deposit its pollutants in harmless quantities across the face of the earth. But when contamination becomes so extensive that water is no longer able to disperse the poisons that it carries, then all life is threatened. Much of the world's water is in such a state today because of the size of human populations and because of the carelessness with which water is misused.

Less than 1 percent of the world's total water supply is available for the nourishment of land plants and animals. The rest is either ocean salt water or ice locked in the polar ice caps. The world's relatively small supply of fresh water circulates perpetually through the atmosphere, the upper layers of the lithosphere, and the living processes of the biosphere. Evaporated water collects in the atmosphere, falls as rain, then passes over and through the land and living plants and animals to return to the oceans in a perpetual pilgrimage called the hydrologic cycle. Water is a medium of exchange among the spheres of life, for at each station in its cycle it both gives and receives.

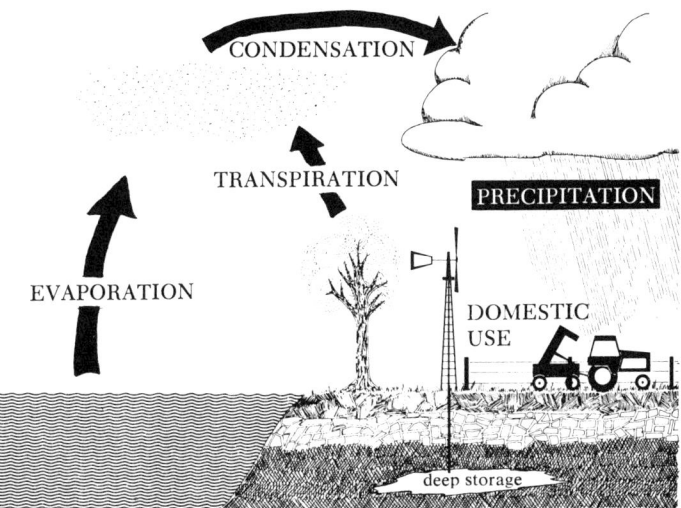

Water passes among the spheres of life in a complex path known as the hydrologic cycle. Alternately a solid, a liquid, or a gas, water readily changes its state and its location. Water is thus a medium of exchange throughout the ecosphere, and the key element supporting life on earth.

THE SPHERES OF LIFE

Raindrops or snowflakes often form when water vapor collects around tiny particles in the atmosphere. If these particles are mere airborne dust they will be returned to the earth with precipitation, leaving the air cleaner because of water's passage through it. But if they happen to be particles derived from fuel combustion, chemical vapors, or radioactive elements, precipitation will spread their dangerous contents over the land. Thus atmospheric pollutants produced in large cities are often found in the snowfall of remote wilderness areas. In recent years, no snow sample collected in North America has been free of lead particles derived from automotive fuels. In some parts of northern Europe, industrial pollutants are so abundant that black snow often falls.

Once it has reached the earth's surface, water performs a dazzling variety of tasks, fetching and carrying across the land. If it falls where there is no soil, it can freeze in the crevices of rocks, splitting them into smaller pieces and thus beginning the process of soil formation. Or perhaps it will collect upon a glacier which grinds mountainside rocks and carries them to valleys below. Any loose earth is likely to be moved by water to lower elevations.

Flowing water carries loads of matter downhill for deposit upon lake bottoms or in the sea. Layers of water-borne sediments deposited in this manner over many millions of years constitute a large

The Hydrosphere

part of the land surface upon which the processes of life depend. When human activities add toxic substances to the moving waters of the earth, their effects will be evident somewhere. Sewage, industrial wastes, and agricultural chemicals are treated by water the same as any of the harmless or beneficial materials that it carries. But instead of creating new environments to support life, these additives may accumulate in water to limit or destroy life. The dead fish and dead plants in today's polluted waters testify to mankind's misuse of the surface portions of the hydrosphere.

Because this planet has a water supply, it has life. Ancient mythologies and modern science agree that life originated in water. Every living creature, no matter how simple or complicated its form, depends upon water for its existence. Water is as necessary to a microscopic bacillus as it is to a human being.

Blood is thickened water, as are the many other fluids essential to life. As water passes through living cells and tissues, it often deposits what it has acquired elsewhere in its travels in exchange for the wastes produced by living organisms. Water molecules can collect minerals or insecticides from their passage over the land, deposit these in the fatty tissues of mammals, and then pass on the mammals' bodily wastes to the land. Both plants and animals also return water to the atmosphere

THE SPHERES OF LIFE

through respiration. A stalk of corn, for instance, will absorb some 200 quarts of water through its roots during one growing season, and will return most of it to the air. Water in the biosphere is in continuous motion, passing from cell to cell of living matter, collecting and depositing at every stage.

The effects of water upon the processes of life are discussed in later chapters devoted to the biosphere. For now, it is enough to note that without water there would be no biosphere, and that the chemical contents of water determine the kinds of influences it will have upon living processes.

Most of the world's fresh water is located beneath the earth's surface, stored in underground lakes or moving among layers of soil and rock on its return to the sea. Groundwater repeats in the earth the familiar pattern of exchange already described, depositing what it has collected elsewhere and acquiring new elements from the layers through which it passes. It has often been said that water is purified as it passes through soil and rock, but that is only true if the chemicals deposited there are more dangerous to life than the ones acquired. If the earth is poisoned, its waters will be poisoned also.

The ancient discovery that underground water can be brought to the surface for use by plants and animals has made it possible to support life in otherwise arid places. As human populations have grown and their technology has increased, it has

The Hydrosphere

become possible to pump great quantities of water out of the earth and to affect drastically the subterranean processes of the hydrosphere. Beneath densely populated or heavily irrigated areas, water tables have been seriously lowered and rainfall is often insufficient to replenish them as quickly as they are used. Ecologists now speak of "water deficits" in those parts of the earth where subsurface waters are disappearing because of human overuse. When groundwater is brought to the surface, much of it inevitably flows off to the sea without returning to the earth's deep fresh water reservoirs.

Water is indifferent to life, yet its quality is reflected in every living thing. Clear water has long symbolized both purity in nature and the spiritual purity of human beings. When water is murky or poisoned, as so much of it is now, human errors have made it so. If humans accept a responsibility to encourage life, then clear and abundant water must stand at the top of their priority list. For as the hydrosphere is, so is life on earth.

4.
The Atmosphere

Why is there air? The question is neither silly nor irrelevant to practical life. The existence of air is another of the unique conditions that makes this planet biologically habitable. But life does not come from the atmosphere; rather, the atmosphere comes from life. There is air because there are plants and animals.

To be more accurate and more specific, green plants produce oxygen, the ingredient in our atmosphere of greatest interest to humans and to other animals whose lives require it. Oxygen is a byproduct of photosynthesis, the process by which green plants convert solar energy into chemical energy. For millions of years plants have been releasing oxygen into the atmosphere, where it is stored and dis-

THE SPHERES OF LIFE

tributed more or less evenly around the earth's surface.

The other major ingredient in the air that we breathe (some 80 percent of it) is nitrogen. Nitrogen is produced by many complex processes, including bacterial decay of plant and animal wastes and volcanic eruptions. Even the discharge of atmospheric electricity by lightning bolts plays a role. The lungs of humans and other air-breathing animals cannot perform properly without a balanced recipe of oxygen and nitrogen, plus just a pinch of carbon dioxide. A few other gasses are also needed in tiny amounts.

Since the atmosphere is produced by ecological processes, it accurately reflects the conditions of life on earth. If the numbers and kinds of green plants growing on earth are significantly altered, atmospheric oxygen will vary accordingly. A mature deciduous forest releases abundant oxygen into the atmosphere, but when it is cut and replaced by food crops the oxygen-producing potential of the land is sharply diminished. Paved roads and parking lots produce no oxygen at all. The cumulative effect of several thousands of years of "land development" by humans has reduced the world's sources of oxygen by decreasing the numbers of growing plants.

Marine plants produce more of the world's oxygen than land plants, but this supply, too, has been diminished by human influences. Pollutant

The Atmosphere

chemicals in the sea may destroy marine phytoplankton or reduce their capacity to produce oxygen through photosynthesis. Many pollutants also kill or disrupt the bacterial actions in the soil or the sea which release nitrogen into the atmosphere.

Unaware of the effects of their actions, humans have significantly reduced the sources of breatheable air. Recent estimates show that the United States now consumes almost twice as much oxygen as its growing plants are capable of producing. Perhaps the more abundant oxygen-producing plants in other countries may help to reduce that deficit. If so, then surplus oxygen may be the largest unpaid export of those countries to the United States. It must be remembered, however, that the same causes which have reduced oxygen production in the United States are also to be found throughout the world.

When fossil fuels such as coal or oil are extracted from the earth and burned, they consume oxygen and they quickly transfer some of their chemical contents from the lithosphere to the atmosphere. Collections of particles from smoke and automobile exhaust are now common over cities and industrial areas. Special atmospheric conditions above some cities like London and Los Angeles tend to trap pollutants and concentrate their effects, sometimes creating major health hazards. Eventually, weather changes always break up these con-

THE SPHERES OF LIFE

centrations and spread their pollutants over much larger areas for all to share. The entire globe is now surrounded by a layer of polluted air that can be seen by jet travelers and measured by scientists. Atmospheric pollution is a world phenomenon, not merely a local problem.

When chemical pollutants are breathed in, they often replace oxygen in the blood and cause strain upon the heart and lungs. Air pollution is known to be a major contributing cause of increasingly common respiratory diseases and heart ailments. Rural residents are less likely to suffer from pollution-caused diseases than city-dwellers, but no place is entirely free of air pollutants. The atmosphere is an excellent system of distribution, which assures that air pollutants will be dispersed among all breathing creatures.

Polluted air attacks more than hearts and lungs. Plants and trees are often deformed or killed by the tainted air surrounding them, and even steel and stone are etched and eroded by airborne industrial chemicals. Many works of European sculpture and architecture survived wars, weather, and vandalism for hundreds of years but have now been defaced by only a few seasons of exposure to chemically polluted air.

People talk about the weather and they do a great deal to change it, often unintentionally. When trees are removed from large areas of land, surface

Pollutants collect in layers of the atmosphere and spread widely around the globe. Their suspended chemicals are harmful to many forms of life. Pollutant layers sometimes filter the sun's rays, changing temperature balances and growing conditions on earth.

THE SPHERES OF LIFE

wind patterns in the local area are usually modified. Downwind from deforested areas, moisture content may be reduced because of the absence of water vapor transpired by trees, and both decreased rainfall and greater extremes of temperature may result. Manmade lakes also change the weather patterns over the land surrounding them.

More significant changes in weather and climate may be caused by various forms of air pollution which influence the temperature of the earth as a whole. The atmosphere regulates heat from the sun, reflecting much of it but holding enough heat so that relatively moderate temperatures are maintained on the earth's surface. Astronauts who have visited the thin atmosphere of the moon with its extreme temperature ranges know well how important earth's atmosphere is to the maintenance of even temperatures.

Atmospheric pollution from smoke, auto exhaust, and jet airplanes appears to influence the earth's average temperature by increasing the "greenhouse effect" of the atmosphere. Even a small increase in the earth's average temperature (1 or 2 degrees) would lead to major consequences. A slight increase in the melting rates of polar icecaps, for instance, would cause a significant rise in the level of the oceans and major changes in continental weather patterns. The climate of the earth depends upon a delicate balance of atmospheric conditions

The Atmosphere

and even small manmade changes in this balance can have far-reaching effects.

Looking up at the sky has always been psychologically and spiritually important to mankind. Clear air permits long views of stars, distant horizons, and goals to be reached, while dirty air has a numbing and depressing effect upon those who must breathe it and squint through it. The state of the atmosphere influences not only the biological conditions of life, but also the state of the human mind.

A wise man once said, "Where there is no vision, the people perish" (Proverbs 29:18). He meant spiritual vision, but this is an extension of the human ability to see clearly the surrounding world and to appreciate its intricate processes. The sky has long been symbolic of powers transcending the lives of individuals, giving people a sense that their existence has more than mere transitory significance. To the extent that human activities have degraded the atmosphere, they have also decreased our ability to find meaning in the sky.

5.
The Biosphere: Plant Life

Long before the simplest animals appeared, thousands of species of highly developed plants had established their family lines and had formed complex living communities. The study of plants is the study of the roots of life, and the history of life on earth is essentially botanical history. The future prospects for life will depend largely upon what happens among the plants.

Many millions of years ago, green plants hit upon a unique process called photosynthesis and have retained exclusive rights to it ever since. The advent of photosynthesis made possible the development of environments capable of supporting diverse and abundant forms of life. Only green plants are capable of photosynthesis, and its importance is difficult to exaggerate for it is the source of most of the energy available to living creatures. All food chains are anchored to plants, the world's primary producers of nourishment. Foods provide energy that photosynthesis has stored in plant cells. The energy in fossil fuels was collected by ancient living plants which were later converted by geologic processes into coal or oil. The oxygen that makes our atmosphere breatheable is another product of photosynthesis. Without the environmental influences provided by green plants, only a few rudimentary organisms could exist on earth.

Plants are also the keepers of soil. Root systems bind loose earth and hold it in place, and provide micro-environments which encourage the chemical and bacterial processes necessary to soil enrichment. Plant wastes and the decay of dead plants add further to the fertility of soils. Virtually every phase of plant life contributes necessary elements to living environments. Plants receive from the sun, but they give to all other forms of life.

Newly exposed surfaces on the earth, such as those left by retreating glaciers or created by fresh volcanic deposits, are likely to be colonized first by pioneering plants. Small examples of such ecological pioneering can be observed following forest or range fires or around the edges of construction projects, where plants like fireweed or dandelions are normally the first to grow. These pioneers are capable

The Biosphere: Plant Life

of living where other species could not survive, but they represent the early stages of preparation for the more complex forms to follow. Most of the plants called weeds are tough pioneering species, which is one reason why gardeners and farmers find it difficult to get rid of them.

Plant succession is the ecologist's term for the natural process leading from simple groupings of plants toward complex botanical communities. Pioneering plants, like human pioneers, are capable of surviving under harsh or inhospitable conditions. Their influence, however, makes the land more suitable for the delicate and specialized kinds of plants which will appear in later stages of ecosystem development. As they live and die, pioneering plants make possible the more complex life forms to follow.

The process of ecological succession begun by the pioneering species, if left alone, leads toward a mature or "climax" ecosystem. Climax communities are extremely diverse and complicated groupings of living plants existing in a relatively balanced state with one another and with their nonliving environment.

In a mature spruce forest, for example, thousands of highly specialized types of bacteria maintain stable soil chemistry as each type plays its particular role in the processes of decomposition. Insects live upon plants and are in turn eaten by birds and other animals; small mammals breed in the complex

THE SPHERES OF LIFE

vegetation; larger mammals eat certain specific kinds of plants or prey upon smaller animals. The many highly specialized plants, from small ferns to enormous trees, make up the setting for all other life, provide food and shelter, and in turn depend upon the environmental determinants of weather and geography. It is an unbelievably complicated community in which no individual and no species can survive well unless many other species also prosper, for all are ultimately dependent upon the completeness of the environment as a whole. The diversity of a climax ecosystem is one of the secrets of its durability.

Some 10,000 years ago, humans discovered how to manipulate plant ecology for their own advantage by the use of agricultural techniques. Farming is a system designed to arrest natural plant succession and to encourage the rapid and abundant growth of the few plant species which provide human food supplies. Farms are ecosystems maintained at a pioneering stage of plant succession, and farmers must work hard to prevent the natural patterns of ecological development which always tend toward greater botanical complexity.

Imagine, for instance, two ecological inventories taken on the same acre of prairie woodland, one a century ago and the other today. The forest of 100 years ago would include several hundred different plant species, many species of soil organisms,

Plants encourage the development of fertile soils. Simple plants help to decompose bare rock and to retain nutrients on its surface. Later stages of plant succession further break down rock into its components and help to establish deeper soils. Eventually, complex communities of plants and animals flourish upon land that was once barren.

THE SPHERES OF LIFE

and a complex population of insects, birds, and mammals. Occasionally a human, perhaps an Indian, might pass over that acre and take from it enough meat or berries to feed his small family.

An inventory of the same acre today might reveal that only one plant species, an annual grass called wheat, grows there. Other plant species have been eliminated by logging, burning, spraying, or cultivation. The bacterial diversity of the soil is also much less than it was, and the soil now requires supplements of fertilizer and nutrient chemicals to support the single plant species that is grown in it. The numbers and kinds of insects, birds, and mammals have been reduced by perhaps 90 percent or more. Now and then a human, riding a tractor, passes over the acre and takes from it enough food for thirty or forty people.

Ecological diversity, and the consequences of mankind's habit of simplifying ecosystems, will be discussed in chapter 11. Increased human dependence upon plants for food has led to major changes in the botanical order of the world. Without these changes, the human population could not have grown to its present size, but with them, the stability of plant ecology has been endangered. It is not yet known how far the simplification of plant life can be pursued without weakening the botanical basis for all animal and human life.

Plants take root in the human imagination as

The Biosphere: Plant Life

they do in the soil. Cultural symbols as old as human consciousness associate plants with fertility and the renewal of life. Flowers appear at both weddings and funerals, and green, the color of photosynthesis, is also the color artists and designers prefer to use when they want to create impressions of human peace and happiness. Humans seem to have always known that their best interests and highest hopes are intimately connected with botanical life. An ecological understanding of plants does not contradict that awareness, but deepens it.

6.
The Biosphere: Animal Life

About 1 million different animal species live in the environments provided by the earth. Large though that number may seem, it is less than 1 percent of the total number of species so far produced by evolutionary history. The overwhelming majority of animal species have proved to be unfit for life and are now extinct. Animal life is precarious, for its continuation depends upon subtle and changeable ecological circumstances. The ability of an animal species to adapt itself to environmental change is perhaps the most important factor influencing its prospects for survival.

Animal species, like plant species, appear in ecosystems according to stages of ecological succession. Pioneering animal species are those best able

THE SPHERES OF LIFE

to survive in sparse or degraded environments. Some insects are well suited to life in burned-over forests, and such adaptable animals as rats and magpies are able to make their livings in lean environments where more highly specialized creatures would starve. As plant succession proceeds and environments become more complex and stable, specialized animal species can begin to make their homes in them.

Sophisticated predators like wolves or eagles require rich and dependable environments if they are to survive and prosper. Such climax animals are not "kings" over their ecosystems, but loyal subjects. Their existence is fragile, for it depends upon a delicate balance among the plant and animal species with which they share their surroundings. When mature ecosystems are disrupted by natural or man-made changes, the most highly developed animals are usually the first to suffer and the first to disappear. Degraded ecosystems are places where rats can thrive but wolves cannot.

Because most animals are more mobile than plants, they need not remain in the same environment all their lives. If food is scarce in one valley, it may be abundant a few miles away. Migrating animals like salmon, caribou, and ducks travel great distances to reach seasonal food supplies or mating grounds. Animals are seldom dependent upon the ecological stability of a single geographic area, yet

A wolf clan, like a human family, is bound together by mutual cooperation and affection. Disruption of normal animal relationships because of human hunting, domestication, or modification of habitat often threatens the welfare and the survival of wild animals.

THE SPHERES OF LIFE

they require a relatively healthy world ecology if they are to survive.

Animals in ecosystems may be described according to what they eat and what they are eaten by. The energy produced by plants is the basic source of nourishment for all animals. Herbivores are those species that consume plants directly, and they are by far the most abundant in nature. Insects and moose munch leaves, rodents and birds eat seeds, and whales swallow great gulps of plankton. Most animals have found that the best way to acquire energy is to eat the plants which produce it.

As long as plant ecosystems are healthy, the animals dependent upon them can remain healthy also. When pollutants are present in plants, they will generally be passed along to animal consumers. And if certain kinds of vegetation disappear, then those animals unfortunate enough to be heavily dependent upon them will have to disappear as well. The caribou lives mostly upon lichens and must go wherever these plants are to be found, but if the lichens have been contaminated by nuclear radiation or destroyed by fire, then the caribou is in serious trouble. All things considered, however, herbivorous animal species enjoy good prospects for survival compared to the carnivores.

Eating other animals, as carnivores do, is in some ways less efficient than eating vegetation. An eagle eating a mouse can benefit from only about

The Biosphere: Animal Life

10 percent of the energy which the mouse has acquired from the vegetation it has eaten. If a steer needs a bale of hay to make a pound of steak, then a human could gain ten times the energy by eating the hay rather than the meat. At each higher feeding level on a food chain, energy becomes more highly concentrated and is transferred less efficiently from one animal to another. Yet the highly concentrated diet of carnivorous animals makes possible their more vigorous ways of life.

Since carnivorous animals are at or near the top of their food chains, their survival depends upon the completeness of lower food links. When rabbits die off from starvation and disease as they periodically do, then the lynx, which depends upon rabbits for most of its food, must necessarily decrease in numbers. Songbirds requiring particular kinds of insects for food cannot survive if insecticides have eliminated their prey. Specialized and highly developed meat-eaters occupy the most delicate of all ecological niches, and they are the first kinds of species to disappear when ecosystems are disrupted.

Omnivores, animals with more generalized dietary needs, have a better chance to survive environmental changes. If garbage from human settlements happens to be the most abundant food supply in an area, then omnivorous animals like bears, gulls, and ravens are not likely to starve to death. To be sure, they will run high risks from contaminants

THE SPHERES OF LIFE

and inedible material in such food sources, and from little boys with air guns and big men with shotguns. But in the long run, perhaps those animals capable of eating bizarre and polluted foods may inherit the earth, for such foods are increasingly abundant. Even so, no animal, regardless of what it eats, can survive in isolation, for each is dependent ultimately upon many other species of plants and animals which make up ecological food chains.

A few animals live at the very top of food chains where they derive their energy from many plants and animals at lower levels, but are not normally eaten in turn by anyone higher up. These terminal species—such as wolves, eagles, sharks, and perhaps humans—normally return their accumulated energy to the food chains from which they have acquired it by recycling their wastes and their bodies. Higher animals can contribute energy to ecosystems by returning their chemical substances to the earth, where soil organisms will begin the whole process over again.

For most animals, the period between birth and death is filled with varied possibilities. Choices and alternatives are always present to some degree. Some choices are rather strictly governed by instincts which have evolved with the species over long periods of time. Among the many possible choices open to an animal species in the course of its life, some kinds of normal behavior must be

The Biosphere: Animal Life

established and regulated if the species is to survive.

From the insects to mankind, animals live according to forms of social organization which have evolved and have adapted the behavior of each species to its appropriate ecological environment. Every animal either is born with or learns standard forms of behavior for such important life activities as the transition from adolescence to adulthood, the acquisition of food, mating and reproduction, the rearing of offspring, family relationships, defense of territory, and the organization of groups of animals into working social structures. These patterns of behavior change and develop as animals evolve in response to their changing environments.

Humans influence the survival of animals by hunting, by domestication, and by modifying the ecosystems upon which animal life depends. Many species have been hunted to extinction because mankind has not accepted the restraints typical among other hunting predators. Good hunters, whether they are wolves or men, must take care that their prey species will not disappear from the earth. Animal domestication is one way humans try to assure themselves of a continued supply of meat, but domestication causes significant changes in the physical characteristics and in the behavior of animals and deprives them of their natural identity. The most devastating effect of mankind upon animal life has been caused by man's modification of

THE SPHERES OF LIFE

the natural ecosystems necessary to animal survival. Environments which have supported animal life for millions of years have fallen to the bulldozer, to pollutants, and to the sheer weight of demands made by an overpopulated humanity.

People sometimes ask why they should care about the welfare of animals. One good reason is that animals can remind us of what we have forgotten: how to live within ecological limits. By studying animals, it is possible to learn how to live in better harmony with the natural world. Animals have a right to life that is as firmly founded as man's, and in most cases more deeply rooted in the past. Finally, the ecological conditions necessary for the survival of wild animal life are the same as those required for continued human survival. Whatever harms animal life is dangerous to mankind as well.

7.
The Noosphere

The word "noosphere" was coined by the French scientist and theologian Pierre Teilhard de Chardin (1881–1955) to designate the unique group of functions that are characteristic of the human mind and spirit and distinguish our species from all other animals. Many animals are capable of communicating among themselves, and they can learn from experience, adapt to varied environments, and pursue highly complicated forms of ritual behavior. Only humans seem able to create symbolic systems such as language and mathematics, to record and remember events over long periods of time, to imagine things that are not present, and to create elaborate systems of culture and belief which can be passed along from generation to generation. The

THE SPHERES OF LIFE

noosphere is the combined and cumulative product of these unusual human activities.

Written records of the noosphere exist for only about 6,000 years of its total history. Since the human species is at least 1 million years old, this means that there are records of only a tiny portion of the human past, less than 1 percent of it. Archaeological and anthropological evidence confirms that our species has changed little in form over the past 250,000 years. During most of that long period, man probably lived as a hunter who was as completely unified with his ecological environment as any other hunting animal. During the past 6,000 years, the period of rapid development of the noosphere, mankind has become separated from natural ecology and has acquired the ability to influence the environment in significant ways. It remains to be seen whether the noosphere can become compatible with the lithosphere, hydrosphere, atmosphere, and biosphere.

Humans loosened the links of their ecological food chains about 10,000 years ago when they learned how to control the reproduction and growth of plants and animals. They found that animals could be domesticated and raised in enclosures and need not be hunted in the wild. Equally significant was the discovery that plants could be grown in concentrated areas for exclusive harvesting by humans alone. Animal husbandry and agriculture were

The Noosphere

early products of the noosphere, and are probably the cornerstones upon which it rests. They have made human civilizations possible, and have led to the unique forms of mentality and behavior which are now mankind's most distinguishing features.

The ability to control certain species of animals and plants has encouraged the belief that nature is mankind's private property. This belief is evident in many religious and mythological traditions which originated soon after the development of agriculture. Ancient Greek and Hebrew stories of creation describe the first stages of human life as if the earth were a garden or small farm created especially for the benefit of mankind. In these traditions, mankind is assumed to have precedence and power over all other forms of life. It has never been clear whether "dominion over nature" means that mankind is obligated to protect and to maintain the natural systems of the earth, or is free to exploit them for exclusively human purposes. Too often, mankind has acted as if the natural world were evil or depraved and has devoted much effort to the control or destruction of nature in the belief that overcoming biology is equivalent to overcoming evil.

In the noosphere, rights are often decreed by status. Since nature has been judged to be spiritually inferior to mankind, no ethical restraints have prevented its exploitation or destruction for human advantages. Natural ecosystems enjoy none of the

THE SPHERES OF LIFE

rights or protections ensured by our laws and customs to human beings. When large modifications of natural systems are planned, such as the clearing of forests or the damming of rivers, only those consequences serving human purposes have been considered, with little or no thought for the effects such projects will have upon other forms of life or upon the ecosystems which sustain life. As the power to modify natural environments has accelerated in the twentieth century, the effects of such actions have been magnified. Where once only local ecosystems were modified, now human actions have begun to influence the entire world ecosphere.

Human pride has often been based upon the belief that mankind is at the center of the universe, that we are the most important creatures to exist, and that we have the power to use or destroy whatever is not human without fear of punishment or loss. We deserve this special status, we have believed, because the human mind and soul give us a destiny higher than that of all other creatures. Gradually, these beliefs have been contradicted within recent history, and they now seem to have lost much of their former power over human thought.

In the sixteenth century it was proved that the earth is not at the center of the universe, but is a relatively modest planet orbiting around an unremarkable star far from the center of an enormous galaxy. In the nineteenth century the great

Images and ideas in the human brain are powerful influences upon natural forms and processes. Modern man is in a maze of his own making, constructed by a cultural tradition which has led to the manipulation of nature for exclusively human purposes.

THE SPHERES OF LIFE

English naturalist Charles Darwin (1809–1882) taught us that human life is connected firmly to all other animal life and that the human species has evolved slowly just as all other mammals have. Traditional religious beliefs have recently lost much of their former power as the world has come to seem less orderly and less benevolent. And early in the twentieth century psychology demonstrated the power of emotional and subconscious forces in the human mind which are largely beyond conscious control. The acceptance of such views has been damaging to many traditional ideas about human dignity, for all suggest that man is not the master in his house.

The most recent blow to the human ego has been dealt by the ecological crisis, for it has made us aware that much of our human pride has been achieved by destroying the natural environments upon which our biological existence depends. Now that we seem to have almost succeeded in conquering nature, long a goal of our cultural tradition, we have begun to discover that this impact upon natural processes is dangerous to human welfare and to prospects for the future. If the triumph of the noosphere can be achieved only at the cost of destroying the world ecosphere, then it will be a hollow and self-defeating victory.

Environmental problems cannot be solved merely by a few technological adjustments. Human

The Noosphere

uses of nature have been guided by religious and philosophical beliefs and by the thoughts and feelings nurtured by our cultural tradition. Until significant changes occur in the human values and beliefs leading to the use of our powers in ecologically destructive ways, there can be little relief from environmental crisis. The most elaborate technology is unlikely to invent an anti-pollution device to cleanse the human mind of the thoughts and feelings which have led to the brink of environmental disaster.

The science of ecology is the noosphere's most recent attempt to understand the other great spheres of life and to seek new kinds of integration between human life and other forms of natural existence. As a science, ecology is a growing body of knowledge of the essential interrelationships among the various systems and organisms which constitute life on earth. But ecology also implies a new way of thinking about the world and mankind's place in it. It suggests that humility before the earth is a better guide for human behavior than excessive pride, and that changing ourselves to agree with the world is a higher goal than changing the world for our own pleasure.

The human body does not adapt rapidly to environmental changes. Our lungs will not learn how to breathe smog, our nervous systems will not remain stable in the midst of noise and overcrowding, nor will our digestive processes remain healthy

THE SPHERES OF LIFE

when they are infused with pollutant chemicals. Like most other mammals, our species is too complex and reproduces too slowly to adapt promptly to new dangers.

The most effective system of change is perhaps the noosphere itself. If we cannot change our bodies, perhaps we can change some of the beliefs and values which have always guided our behavior. The human mind has always before found ways to think differently when it was necessary to do so in the interest of human survival. Ecology challenges us to discover how to adapt the values of the noosphere to the requirements of a whole earth in which mankind will persist as a valuable and contributing species.

8.
Paths of Energy

If energy and resources were used by human economy as they have always been by natural ecology, it would be possible to avoid energy crises, food shortages, and most serious pollution. Many of the environmental threats facing mankind have arisen from needless wasting of the raw materials necessary to life. Humans have acquired the disastrous habit of taking from the earth without offering anything in return. As Kenneth Boulding, an American economist, has put it:

> Ecology's uneconomic,
> but with another kind of logic,
> economy's unecologic.

THE SPHERES OF LIFE

Though the economic costs of developing sound energy sources may be high, the failure to do so is sure to be catastrophic.

Energy cannot be created or destroyed. It may, however, occur in concentrated or dispersed states. A candle is a relatively concentrated form of energy. If it is a tallow candle, the energy it contains was once stored in the fat cells of an animal, which in turn received its energy from the plants it had eaten, which in turn had converted the sun's energy to chemical energy through photosynthesis. At each stage in this complicated process, the original energy from the sun passed through less and less concentrated forms. The end of the road is reached when the candle is burned and most of its energy is dispersed through the room as heat and light. There is no way to reassemble those products in order to make a candle again. Once energy becomes heat, it is unavailable for use by living creatures.

Paths of energy through natural process keep as much energy as possible moving through the systems of life. If the fat of an animal is eaten instead of being converted into a candle, some of its energy will be used to maintain another form of life. The economy of ecosystems, called the food chain, keeps maximum amounts of energy flowing through living organisms. The only significant source of energy is the sun, and the only way energy can be lost to life is through its dispersal as heat.

Once the fuels of the earth are burned, their stored light and heat become forever unavailable. Civilization has burned its candle flame brightly in recent centuries, consuming fossil fuels and radiating much heat and light. But the wax and wick of the world are limited and cannot be burned forever.

THE SPHERES OF LIFE

Human economy is unecologic when it accelerates the conversion of energy sources into heat. That is what most modern economies do best. Resources are extracted from the earth and converted by industry into products which may be sold. The smoke from industrial chimneys is one indication of the amount of heat that is lost in most such manufacturing processes. And many of the items so manufactured are themselves machines which consume energy and radiate more heat, such as motor vehicles, electrical appliances, farm and industrial machinery, and all of the mechanized gadgetry of modern life. However much wealth and comfort such an economy may produce, its most abundant product is heat.

Fossil fuels resemble a candle in that they are concentrated forms of energy derived from once-living organic matter. When oil is pumped from the earth for burning by automobiles and industries, its energy is forever lost. That is why coal and oil and other such minerals are called "nonrenewable," for once they have been transformed into heat their energy cannot be reused by any form of life.

Man's mastery of fire may have been the first step toward civilization and culture, but man's first fire also lighted the candle of resource depletion. That candle has illuminated the human way for several thousand years, but it has begun to sputter

Paths of Energy

in the twentieth century as its fuel has been more rapidly consumed. The size of the flame has been increased so that it gives off enormous amounts of light and heat, but no way has been found to reconvert the heat into a candle again. When the wax and wick of the world are used up, there is no corner store where we can go to buy another such candle.

The sophisticated technology of recent decades has found many new ways of burning the sources of energy, but it has also lately begun to discover methods of conserving them. These methods are largely human attempts to imitate the principles of energy conservation which have always been characteristic of natural ecosystems. Perhaps the time is near when humans can apply these principles to their economic life.

When green plants convert solar energy into chemical energy through photosynthesis, they do not consume nonrenewable resources or pollute their environments. Human power plants cannot so far make the same claim. Only recently have ways been found to imitate the green plants by converting solar energy into electrical energy without the use of fossil fuels or water power. The technical problems of producing electricity from solar energy are solvable, but so far economic customs have not encouraged the development of this hopeful new power source, perhaps because many jobs, indus-

THE SPHERES OF LIFE

tries, and fortunes depend upon the traditional exploitation and destruction of energy resources.

The sun seems the most likely source of energy which does not require the depletion of nonrenewable resources, but there are also others deserving of further exploration. Geothermal energy from natural heat stored beneath the earth's surface has already been successfully used to produce electricity in many parts of the earth. The winds and ocean waves may also become sources of power if we can learn to use their energy properly, and neither requires the destruction of resources, increased pollution, or high heat loss. Until such alternative sources of renewable energy replace the burning of fossil fuels, we will probably remain trapped in the present futile patterns of growing economies dependent upon diminishing resources.

Petroleum deposits are the organic remains of plants and animals which lived millions of years ago. Once they were edible, and they can be again. Scientific techniques are now available for converting crude oil into food protein. Known microbes like yeasts and bacteria are capable of digesting oil and producing almost pure protein from it. Technology is also available which can transform this protein into foodstuffs palatable and nourishing to human beings or domestic animals. Enough crude oil is stored in the earth to feed the entire human population for many centuries. It has been estimated that

Paths of Energy

the amount of petroleum now used as fuel in one year would be sufficient to feed 54 billion people for a year, or sixteen times the present world population.

If oil can be eaten, much of its energy will be returned to the life cycles from which it came, but if it is burned as fuel that energy will never again be available. Perhaps we must face the fact that food is more essential to human welfare than automobiles and factories. The hungry people of the world can get along without more cars, more hydrocarbons in their air, or more lung disease. As they increase in numbers during the coming years, surely they will ask what has become of the enormous deposits of petroleum food. If we must answer shamefully that we burned it, it will not help much to add that we burned it by mistake.

Recycling means more than collecting bottles and cans for return to the grocery store. Such habits are necessary to amend our wasteful ways of living, but they are insufficient to solve the problem of diminishing resources. The wasteful misuse of the world's resources will not be controlled until we can learn to provide for ourselves without depleting the capital of energy that natural processes have stored in the earth.

Other forms of life have prospered for millions of years by living according to ecological principles of energy use. Now that we have begun to

THE SPHERES OF LIFE

learn how these principles work, we have an obligation to apply them to human economic practices and to develop technologies which will be consistent with them. The ability to live within the energy limits of nature is available to us if we can find the sense and the will to make human economy agree with natural ecology.

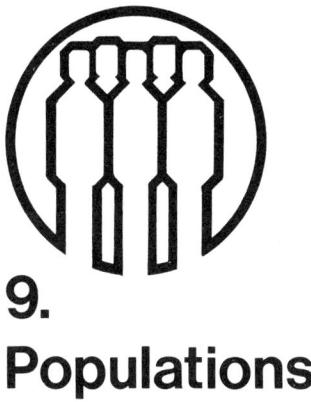

9. Populations

Population control appears to be one of the lesser talents of nature. Every animal struggles to live as long as possible, to reproduce as much as possible, and to postpone death if possible. To the extent that an animal species is successful at these goals, its population will increase. But the increase of numbers in one species inevitably means suppression of other competing species and increased demands upon the environment needed to support life. So the numerical success of a species is commonly the cause of its ecological failure. Rapid and catastrophic fluctuations in animal populations occur frequently in natural ecosystems as the privileges and dangers of high population pass from one species to another.

THE SPHERES OF LIFE

Cancer is a population explosion among cancer cells. If there were chambers of commerce among cancer viruses, they might well applaud their species' rapid conquest of the various territories in the animal body that constitutes their environment. They would not discover until too late that the triumph of the cancer population means the death of the body invaded by cancer. Whether the ecosystem is a single body, a forest, a continent, or the whole world, its stability depends upon maintaining balances among its various populations, and that means preventing the unchecked growth of numbers in any one species.

Population explosion followed by collapse is a normal pattern for some species. Lemmings, rabbits, and many rodents increase and decrease according to predictable cycles. When such populations are at their maximum they rapidly consume their available food supplies and begin to suffer from hunger. Their sparse diet then makes them more susceptible to epidemic diseases and less able to avoid the predators which normally hunt them for food. The numbers of predators also rise with the population of prey species, causing increased hunting pressures.

Under extremely crowded living conditions, psychological pressures cause disruption of normal social structures and abnormal behavior by individual animals. Normal mating behavior is often

Among overpopulated species abundant food does not necessarily ensure peace. When jaeger populations increase in response to an abundance of their favorite food, lemmings, jaegers tend to fight over status and territory more frequently than usual. A full stomach cannot guarantee a peaceful disposition when population pressure is high.

THE SPHERES OF LIFE

affected, and parental care of offspring may become lax or disappear altogether. Overcrowding is sometimes accompanied by physiological changes which cause miscarriage or the resorption of fetuses by pregnant females. Such spontaneous abortions are natural events which play a significant part in population control. These natural controls are the only methods available for regulating populations which have overstepped reasonable limits. Fewer births and more deaths must occur if an overpopulated species is to achieve a size appropriate to its ecological environment.

The problems caused by overpopulation are not necessarily solved merely by increased food supplies. Studies of jaegers, predatory arctic birds which hunt lemmings, reveal that jaeger populations begin to decline when the lemmings are most abundant. Jaegers fight so vigorously among themselves over nesting sites and status that most of them fail to raise normal families. Abundance of food seems related to social disruption and psychological problems which reduce the jaegers' chances for biological success. Apparently wealth does not assure welfare among jaegers, and perhaps not among other species.

High population density can be supported only if extraordinary demands are made upon the resources needed by a species to support its life

Populations

style. North American elm trees have been the victims in recent years of an unchecked population explosion among a small insect, the imported European elm bark beetle, which transmits a fungus that is fatal to elms. With few natural predators in the North American environment, these little bugs have gradually eaten their way across the continent, spreading their fungus and leaving dead elms along the streets of many cities and in parks and forests. Insecticides and control programs may have slowed the beetle somewhat, but it now appears that the population of bark beetles will not be controlled until most of the elm trees are gone. Disaster can confidently be predicted for the bark beetle, and for its favorite resource, the elm tree.

Elm bark beetles live only upon elm trees, but those species with broader needs affect their environments more profoundly when their numbers become excessive. An overpopulation of elephants is capable of destroying its own habitat and that of many other plant and animal species. And the overpopulation of a species like mankind creates environmental demands extending to virtually every component of the world ecosphere. As the human population increases, other species of animals must decline in number, vegetable and mineral resources must diminish, and such secondary effects as pollution and erosion must increase. Overpopulation among advanced species influences the environment far more drastically than increases among simpler species.

Overcrowding is mentally unhealthy for many animals. Antisocial behavior, neurosis and mental instability, violence and cruelty, despair and frustration are frequently found in humans and other higher species living in cramped conditions. Some animals lose their normal instincts for mating and care of their young, resulting in a breakdown of family ties and high infant mortality. Mental balance seems to require a measure of privacy and personal space, plus freedom from the extreme tensions typical of overcrowded communities.

Diseases also spread more rapidly among overcrowded populations. It was no accident that the great plagues swept medieval Europe shortly after large cities were founded there, for epidemics are difficult to prevent in conditions of high population density. Pathological effects of overpopulation, unpleasant though they may be for individuals, all contribute to the solution of population problems. They tend to increase mortality and decrease fertility, and so they lead toward smaller populations in which more orderly ways of life are possible. Nature seems to have provided no painless methods for curing the disease of overpopulation.

No one knows what the ideal population might be for humans or for any other animal species.

Populations

Clearly, populations must be limited by the amount of food that is consistently available. British scientists have recently estimated optimistically that the world can produce enough food for about 3.5 billion human beings on a sustaining basis without severe ecological damage. That population size was passed several years ago. Aside from food, population size must also be judged by its effects upon the ecological environment, by some standards of mental and social health, and by its influence upon human cultural and spiritual creativity.

The world's most respected religions, philosophies, political systems, and works of art originated when the world population was only a small fraction of its present level. High population is unnecessary to the achievement of high human goals, and it may in fact impede seriously the ability to realize human potentials. As there are more of us, the life of each of us becomes less satisfying, less significant.

No matter how sophisticated human technology may be, birth and death will remain the only forces capable of changing population size for humans and for other animals. More births must sooner or later be balanced by even larger increases in death rates. Birth control and high mortality rates are distasteful to contemplate, but one or the other will inevitably be imposed as our species multiplies beyond the ecological limits of the earth. And there are limits.

10.
Evolution

An ecological view of the world became possible when Darwin explained the basic principles of evolution by natural selection. Much controversy has surrounded evolutionary theory since that time, for many people have found this view of life incompatible with other long-established concepts within their cultural tradition. A brief examination of evolutionary ideas may make it easier to understand what ecology implies for our thinking about the world and about mankind's place within it.

The immensity of the earth's time is difficult to conceive for humans whose lives usually span less than a century. If the earth's total history were to be represented by a space as large as the Pacific Ocean, the existence of mankind could be symbolized as a

THE SPHERES OF LIFE

bit of froth on the beach. A tiny fraction of that froth would have to stand for the 6,000 years of civilization's written records, and an individual lifetime would be only a small bubble. Even if only the million years or so of mankind's presence on earth are considered, it is difficult to recognize that the modern industrialized world occupies less than half of one percent of all human history. Our time is a tick or two on the world's clock.

An evolutionary view of time further suggests that the billions of years of earth history did not occur merely to prepare the earth for human use. Time does not proceed in a linear fashion toward human goals, but seems rather to follow irregular cycles favoring different forms of life at different periods. Other species of animals have achieved worldwide importance for long epochs, but now are known only through their fossil records. If there is a pattern in evolutionary time, it seems to favor increasing diversity and complexity of life forms. The clearest difference between life today and life a few million years ago is that there are more kinds of living things. Humanity is one member of a complex world community of living forms evolved over many millions of years.

The evolutionary idea of natural selection among species troubles some orderly human minds because it suggests that the world is governed by chance rather than by design. Natural selection

MILLIONS OF
YEARS AGO

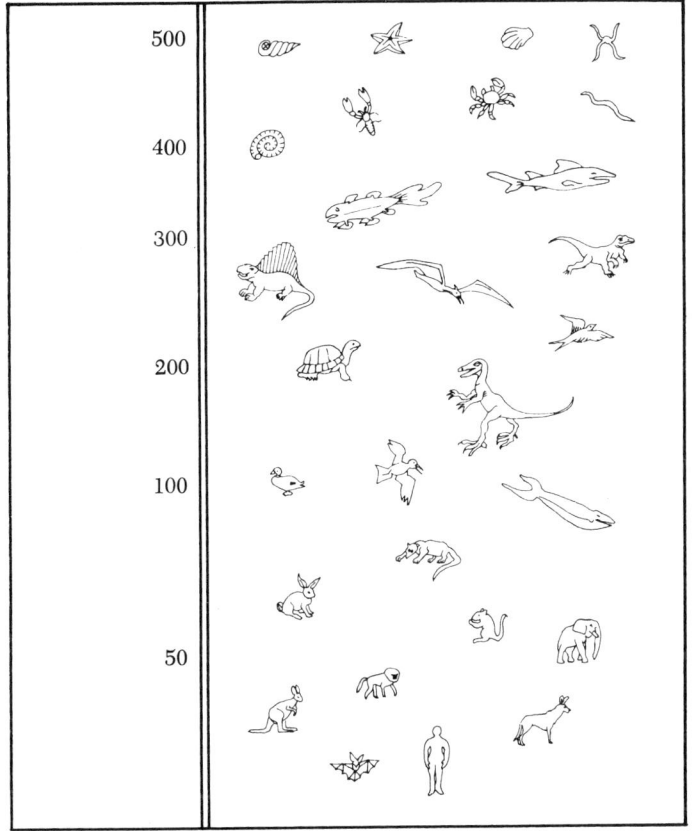

Mankind's million years on earth is brief compared to the long histories of other species. The period of human civilization (6,000 years) is a small fraction of one percent of evolutionary history.

THE SPHERES OF LIFE

means that some species will survive while others suffer extinction, and that the difference between these two fates may be governed by random events which no species can control. The reptiles which once flourished on the earth did not disappear because of errors in reptilian judgment, but because their accustomed forms and ways of life at some point became impossible in a changing world environment. Since the world had changed and their evolution was unable to keep pace, the great reptiles simply became irrelevant and ceased to exist. However much the poor reptiles are to be pitied, it is evident that humans and other species are subject to the same limitations. Nature does not appear to care which species survive, but merely that those doing so must be appropriate and well adapted to their environments.

Organisms evolve new kinds of structure and behavior for better adaptation to environmental changes which might otherwise threaten their existence. Evolution occurs when random variations in form or function help to adapt an organism more perfectly to its changing environment. Natural selection is not determined by battles fought among animals to see which is fit enough to survive and which is not. Rather, the evolutionary process is one of adaptation and accommodation, with the various species exploring their environments in search of

Evolution

opportunities to preserve their existence. Evolution is a matter of muddling through.

Animals have traditionally been regarded either as simple, stupid creatures, or as cruel and malicious destroyers. It seemed no honor, then, when evolutionary scientists announced that humans are closely related to other animals and have evolved from ancestors shared by many other species. The sense of human uniqueness seemed threatened by evolutionary ideas and many people felt deeply insulted by the suggestion that humans are animals, too.

Only recently has it become evident that traditional estimates of animal life have been greatly oversimplified. Recent studies have revealed that the lives of many animals are extremely complex and include most of the same ingredients as human life. Animals, like humans, live according to close family ties. They have regular patterns for courtship, mating, and the rearing of offspring, systems of social order, and methods of land use and territorial defense. The various means by which many animals meet these needs are quite as complicated as human methods used for the same purposes. The family life in a wolf clan is as delicately balanced as that in a normal human farm family, and the social structures of geese serve their species at least as well as politics serves ours. As more has been learned about

THE SPHERES OF LIFE

the intricacies of animal life, it has become easier to accept an evolutionary relationship to animals and to understand that many needs and feelings are shared with them.

However, we apparently do not share with animals our capacity to think. The human brain, with its ability to store knowledge and to understand experience, is evidently a unique evolutionary experiment being tried out on only one species. The human mind has created language, civilizations, technologies, and spiritual and artistic traditions which are not to be found among other animals. It has also created a world environmental crisis, and a science of ecology to cope with it.

The time seems near when it will become evident whether the human mind is to be an evolutionary success or a failure. Evolutionary developments succeed or fail depending upon whether or not they improve a species' adaptation to its environment. Gills would never have been perpetuated among fish, nor wings among birds, if these evolutionary innovations had not proved valuable for survival. The human mind has so far been a wonderful instrument serving the success of our species. It has permitted us to survive and increase in virtually all world environments and to enjoy experiences available to no other animals, but it has also been the source of destructive ideas and instruments which are capable of modifying the world environ-

Evolution

ment and exterminating many species of plants and animals, perhaps including us.

Evolutionary time is long and the history of human consciousness is short. Every organ of the human body has been thoroughly tested over millions of years, except for the enlarged brain which is our most distinctive feature. The survival of our species and of the world's natural environments will probably depend upon what happens within the human mind over the next few decades or, at most, the next few centuries.

If humans persist in the notion that the world must be made to conform to their wishes, then the prospects are dim. But perhaps the mind and its products can be brought into closer agreement with the restraints of natural world ecology. In the end, it must inevitably be determined whether the human brain is to be a continuing feature of evolutionary history or merely a brief episode. Human pride will be fully justified if it can be demonstrated that the thinking mind is capable of participating in world ecology with balance and durability.

11. The Principle of Diversity

Nature is magnificently wasteful. For every dandelion seed that germinates, hundreds more fail to do so. Thousands of years are invested in the evolution of species and ecosystems, only to end in their extinction or degradation. Gigantic evolutionary "mistakes" occur, like dinosaurs whose bulk could not possibly be sustained by the food supply available within their range of motion. In places where one kind of tree or one kind of mammal would seem sufficient, ten or a hundred appear. Though people often speak of ecology as if it meant thrift and careful rationing of resources, natural ecology is in fact prodigal and spendthrift.

Human civilization is wasteful, too, but with a difference. Such wastefulness has typically led to

THE SPHERES OF LIFE

simplification of ecosystems by reducing the numbers and kinds of species living on the land. Civilization exists to serve the needs and desires of only one species, while natural ecology has no bias for or against any particular form of life. Humans squander resources in search of uniformity, while nature recombines its substances into ever more complex and diverse ecosystems.

Though uniformity has many short-term advantages, it often leads to long-term dangers. Identical organisms are subject to identical diseases and misfortunes, and the larger the uniform population the greater may be the misfortune. The corn blight of 1970, which destroyed a significant part of the North American corn crop, occurred after a single genetic strain of corn had been adopted by almost all farmers. A new fungus then appeared which was able to attack that specific strain, and the blight quickly spread from coast to coast. A few decades ago, when farmers chose their seed corn from a wide variety of strains, such rapid epidemics could not easily spread between fields. Dangerous genetic uniformity now exists among many food crops such as wheat and rice, thus increasing the possibility of widespread crop failures.

As the diversity of organic life decreases, technology must be increasingly depended upon to solve the resulting problems. Insecticides can partly replace the natural insect controls once provided by

Natural organisms like the dandelion (*top*) always produce more seeds than can possibly grow. The end result of many such examples of superabundance is a wide diversity of possible combinations. Biological diversity is essential to the stability of ecosystems.

However, human management of biological processes often reduces natural diversity. When complex ecosystems are converted into croplands with a uniform plant life, stability is difficult to maintain.

THE SPHERES OF LIFE

songbirds, and fertilizers can temporarily augment the soils which have been depleted because only one kind of crop was grown upon them. When a fungus threatens corn, scientists can also breed a new specialized variety of corn capable of resisting that fungus. Such efforts, however, always proceed after the fact, for we can never respond with some new technology until we have seen and understood the problem. Thus symptoms are treated rather than causes, and technology must escalate endlessly as new symptoms multiply.

Natural diversity provides in advance for adjustments by the great number of "unnecessary" elements always waiting in the wings for a chance to perform. Some seeds, for instance, are capable of germinating only after they have been exposed to fire; they may rest on a forest floor for decades or centuries before the forest burns, but when it does they are prepared at once to produce new plant growth on the burned area. Similarly, white cells wait in most animal bodies with little to do until a disease organism comes along, and then their numbers multiply to combat it. The disease cured, their population declines to its former modest level. Natural systems include many such idle elements, not because nature is planning ahead for emergencies, but simply because natural systems tend to produce every possible combination of organisms whether they are needed or not. Most, of course, are not, and

The Principle of Diversity

if these remain irrelevant for a very long time they will probably become extinct.

It has been estimated that roughly 1 billion different biological species have existed during the entire history of organic evolution. Ninety-nine percent of that number was "wasted," for only about 10 million species inhabit the world today. The millions of extinct plant and animal species have contributed, however, to the development of the present world environment and to the genetic inheritance of contemporary living species. No one can guess how many animals may have flapped their way awkwardly into oblivion before the first bird with usable wings appeared, or how many amphibians gasped through life before lungs evolved as an effective means of respiration. And who can tell how many different brains it took before one formed the first word? Those species that failed to survive have left their genetic prints upon those now living. Had they been less abundant or less diverse than they were, we would now be less complex than we are.

When human actions reduce the numbers of species on the earth, the prospects for future evolution are diminished by that amount. The list of endangered species now facing extinction because of human influences numbers more than 1,000. That list is made up mostly of mammals, birds, and fish, and it does not include most reptiles, insects, microorganisms, or botanical species. Little is known of

THE SPHERES OF LIFE

how human activities may be affecting the evolutionary diversity of life forms. Where measurements have been possible, the evidence indicates that humans have decreased the genetic diversity of almost every environment they have entered during their recorded history.

Stable systems are able to accommodate changes without serious disruption, and to return to a state of balance even when new elements are introduced. Such stability is possible if a system is complex enough to accommodate novelties. A new species of insect introduced to a tropical forest will have to compete with the many insects already established there, and its influence is likely to be small in the complex life of its new community. The same insect might cause devastation in an agricultural area where there is little competition and little complexity. The greater the diversity present in ecosystems, the greater will be their resistance to disruption by innovations.

Evolutionary adaptation to environmental change requires that a large storehouse of varied life forms must be available from which new combinations can be made. A weed may contain the genetic resistance to disease which will someday prove essential to the survival of a hybrid food grain. Wild songbirds may seem unimportant at one time but may later prevent an insect infestation. As the num-

The Principle of Diversity

bers and kinds of species decline, fewer options are available to respond to environmental changes.

Wilderness land supports a greater abundance of life than tamed land. The preservation of wilderness is extremely important because of the genetic diversity that it produces and maintains. Wilderness is wasteland only in the sense that many forms of life which it supports may not be economically valuable to mankind. The genetic value of wilderness is very high, for it is the world's best source of the biological variability that is essential to species survival and adaptation.

Diversity is a natural consequence of freedom, for humans as for other natural creatures. Uniformity and simplification occur only when concentrated power is exercised by some dictatorial force. As tyrants have occasionally suppressed the natural diversity of human thought and behavior, so mankind as a whole has tried, often successfully, to dictate uniformity to nature. However benevolent such despotism may be, in the long run it is self-defeating, for unified and simple systems have only poor prospects for survival.

Love is a commitment to something that is different from ourselves. Just as males and females love one another because of their physical and emotional differences, not in spite of them, love of nature is a recognition that its abundant life forms do

THE SPHERES OF LIFE

not resemble our own and live according to many and varied processes. Genuine love requires affection and respect for those things that are quite different from ourselves. Love affirms the worth of diversity.

12. Environmental Ethics

Pollution used to mean corruption in the human soul caused by sin. Now it also means infection in natural environments caused by human errors. Both meanings refer to debasement of things that might have remained pure if not for mistaken use of human powers. A polluted environment raises both ethical and scientific problems.

Ethical principles identify the kinds of thought and behavior necessary for a good and just life. Most inherited ethical systems describe how humans should act and think toward others in order to promote order, good will, and love among mankind. Few ethical systems offer much guidance concerning responsibilities toward animals, plants, or the land. Unfortunately, the natural environment

THE SPHERES OF LIFE

has seldom been seen as an important part of our ethical responsibility. An ecological view suggests that human affairs cannot be separated from natural processes, and that a truly good life must provide for the balanced fulfillment of many species.

Virtually every ethical and religious tradition insists that actions should be judged by their effects upon others. We are cautioned to do unto others only what we would have them do unto us. If "others" were interpreted to include "other species" as well as "other human beings," we would be well on the way toward the creation of a new environmental ethic. We would then have to recognize that the endangered species list might someday include our own species, and that the animals and plants now facing extinction because of human actions represent a moral problem of the highest order. Whatever happens to other species reflects what can happen to our own.

The right to life is not an exclusive human prerogative. It belongs to all species, as regulated by natural selection and environmental adaptation. A species earns the right to survival according to its ability to integrate its form and behavior with the changing conditions of its natural environment. The right to life is strongest when it extends equally to many different forms of life. Any animal which exterminates its competitors thereby weakens the com-

The ethical beliefs which govern human uses of nature also direct the way in which humans use one another. When human powers are used for manipulation contrary to ecological principles, the entire world ecosphere—including mankind—suffers distortion and danger.

THE SPHERES OF LIFE

plex web of life supporting its own existence. Success depends upon the integration of species.

Metaphors have influenced the way humans feel their ethical responsibilities toward nature. A metaphor is an imagined comparison between two things intended to express an idea or a feeling. The customary metaphors used to describe the relationship between mankind and the earth usually have suggested that nature is mankind's property and that its features exist only for human benefit and pleasure. Although metaphoric thinking is among the richest and most valuable talents of the human mind, it sometimes leads to dangerous inaccuracies because of the ease with which we forget that imaginative metaphors may distort real relationships.

The most ancient images of man-nature relationships show man as a caretaker who tends a garden earth and reaps the abundant foods that seem to grow especially for human use. Gardens are described as places of moral perfection and artistic beauty, where evil and danger are unknown and peace and harmony reign. Friendly animals, nourishing plants, and the absence of tension or competition are consistent features of garden imagery. These images persist today as strongly as they did several thousands of years ago when the garden metaphor first appeared as an expression of human attitudes toward the earth.

Gardens do not represent nature, but merely

the preferred human uses of nature. Gardens must always be enclosed to exclude some unwanted species; no predatory animals are allowed inside, nor are poisonous plants, biting insects, or tangled undergrowth. When a reptile appears in such a garden, it is equivalent to the entry of sin into a world of moral perfection, as Adam and Eve found out. However appealing the garden metaphor may be, and however deeply it may be rooted in cultural traditions, it is necessary to remember that it does not accurately portray nature. Gardens are intended to satisfy human desires, and they show how humans can use natural processes for their own purposes.

In modern times, a favorite metaphor has likened the earth to a machine designed to serve mankind. Early versions of this metaphor suggested that the earth was a great clock designed to measure the hours of human life. Animals and plants were the springs, wheels, and levers of the mechanism, all harmoniously fitted to keep the world turning while mankind busied itself with higher things. The most recent manifestation of the machine metaphor occurs in the popular conception of a "Spaceship Earth," where the earth is thought of as a finite life-support system for mankind. Spaceship Earth is regarded as a complicated gadget designed to speed mankind on its way toward great adventures and exciting new discoveries.

The machine metaphor, whatever form it

THE SPHERES OF LIFE

may take, conveys the message that nature is a tool in the hands of mankind. Like gardens, machines require human care if they are to be useful for human purposes. Maintenance and proper management must be provided by mankind before nature will yield its benefits to our species. The natural earth, however, is neither a garden nor a machine. There is no good evidence that human pleasure, profit, or progress are of special importance to the world as a whole, or to the many species of plants and animals whose complex lives are lived on the same earth shared by humans.

Ecology is the newest image created by humans to describe their relationships to nature. Like the garden and the machine metaphors, ecology is a product of the imaginative human brain. It is a fresh way for humans to think about the earth and about their role in its processes. Whether it will prove to be more accurate and useful than traditional metaphors depends upon how its meaning is interpreted for ethical values and is applied to behavior.

Ecology implies that humans are animals living in ecosystems as participants, not as possessors. Their first responsibility is to provide for human survival and well-being, just as every other animal seeks to provide for itself and its kind. The surest way to accomplish this is to permit and to encourage the free diversification of other forms of life. Just as survival depends upon adaptation to environment, so stability depends upon an abundance of life forms and great complexity in their interrelationships. A species that is totally involved in its environment is likely to find its own best prospects for the future.

The manner in which humans think about nature is likely to resemble the way they think about one another. If they regard the world as a farm, then they see fellow humans as a crop. They "cultivate" themselves and think of professions as "fields" to be protected from wild influences. Or if nature is regarded as a collection of raw materials, then human beings are considered to be "resources" to be "developed" according to some profitable goal. And when they imagine that the world is a machine, they are likely to treat themselves as mechanisms. Like a rifle they must have "aims," or like a computer they must have "programs."

Powerful people sometimes use their fellow humans just as they use nature. American Indians were driven from the prairies by the same methods and for the same reason that the buffalo were: to eliminate competition for control of the land. The enslavement of defenseless human beings has often followed the same patterns as the domestication of animals. Some "master race" always seems ready to exterminate either people or animals that it thinks to be dangerous, or to tinker with human conscious-

THE SPHERES OF LIFE

ness as if it were a carburetor or a computer. Mankind has formed the habit of conquering other humans by the same means used to conquer nature. But when nature is exploited and polluted, so is mankind. Humanity suffers from the same ills that afflict the natural world.

Environmental ethics require respect for life of all kinds, a knowing acceptance of the necessary processes of nature even when they do not suit human convenience, tolerance for diversity and complexity, and above all the wisdom to adapt to the world rather than trying to modify the world to satisfy human wishes. Humility before the ecological processes which govern all life is necessary for more wholesome relationships between mankind. Good will among men may even be possible if we can learn to live more at peace with the earth.

Glossary
Bibliography
Index

Glossary

ADAPTATION: fitting into a new use or situation, adjusting to environmental conditions
AMPHIBIANS: cold-blooded vertebrates capable of living both on land and in the water
ATMOSPHERE: air and gasses surrounding the earth, and their systems of movement and change

BIOLOGICAL: relating to life and living processes
BIOSPHERE: all the living things of the earth and their relationships

CARBURETOR: apparatus for mixing air with gas vapor in cars
CARNIVORE: a meat-eating animal

THE SPHERES OF LIFE

CESIUM: the most electropositive element known; a dangerous waste product from nuclear reactions

CLIMAX ECOSYSTEM: the stage in ecological development or evolution in which the community of organisms reaches a state of balance and stability

CONSERVATION: the preservation of natural resources, such as topsoil, forests, and waterways

CONSERVATION OF ENERGY: a physical law stating that the total energy of an isolated system remains constant regardless of changes within the system

COSMIC RADIATION: radiant particles from sources of energy outside the solar system

DECIDUOUS: of plants, shedding or losing foliage at the end of a growing season

DISCIPLINE: branch of knowledge; also a system of control enforced by training or obedience

DOMINION OVER NATURE: the exercise of exclusive control by man over the natural systems of the earth

ECOLOGICAL CRISIS: a state of danger caused by disruption of normal relationships between living organisms and their environments

ECOLOGY: the study of the relationships between organisms and their environments

Glossary

ECOSPHERE: the integrated natural systems of the earth viewed as a whole

EROSION: process by which the surface of the earth is worn away by water, glaciers, wind, waves, etc.

EVOLUTION: the process governing the development of species from early forms

FAUNA: animals; particularly the animal life of a particular region or period

FETUS: unborn or unhatched vertebrate

FLORA: plant life; particularly that of a particular region or period

FOOD CHAIN: a series of organisms interrelated in their feeding habits, so that each organism depends for its survival upon the availability of other organisms

FOSSIL: the mineralized remains of a formerly living organism

FOSSIL FUELS: combustible material, such as coal or oil, derived from the remains of formerly living organisms

GEOTHERMAL: relating to the heat of the earth's interior

GERMINATION: the beginning of growth, as in the sprouting of a seed

HERBIVORE: an animal whose diet consists mainly of plant foods

THE SPHERES OF LIFE

HYBRID: a specialized or unique variety of plant or animal, often produced by human breeding experiments with agricultural species

HYDROCARBONS: organic compounds containing only carbon and hydrogen byproducts of combustion

HYDROLOGIC CYCLE: the system governing the distribution and circulation of water on land surfaces, in the soil, and in the atmosphere

HYDROSPHERE: water around and within the earth, and the systems which move and alter it

LICHEN: plant consisting of a fungus in close combination with an alga; lichens often receive much of their nourishment from the air

LITHOSPHERE: the geological and physical systems of the earth, including surface land and deeper layers

MICROORGANISMS: organisms of microscopic size

MILLENNIUM: a period of 1,000 years

MOLECULE: the smallest particle that displays the characteristic physical and chemical properties of a compound

NATURAL SELECTION: the evolutionary principle governing the extinction and survival of biological species

NICHE: the place of an organism in an ecological

Glossary

community, especially in regard to food consumption

NOOSPHERE: human mentality and the ideas, cultures, and civilizations based upon it

OMNIVORE: an animal which feeds on both plants and animals

PHOTOSYNTHESIS: the process by which green plants convert solar energy into chemical energy

PHYSIOLOGICAL: relating to an organism and its bodily functions

PHYTOPLANKTON: minute, floating aquatic plants

PIONEERING SPECIES: animals or plants capable of establishing themselves in hostile or previously barren environments

PLUTONIUM: a radioactive element used as a reactor fuel in nuclear weapons; a radiological poison specifically absorbed by bone marrow

PRECIPITATION: water (hail, sleet, snow, mist, rain) deposited on the earth

RADIOACTIVE PARTICLES: particles emitted in radiation, including alpha particles, nucleons, electrons, and gamma rays

RESORPTION: the act of dissolving and assimilating

SEDIMENT: matter that settles to bottom of liquid; material deposited by water, wind, or glacier

THE SPHERES OF LIFE

STRONTIUM 90: a dangerous radioactive isotope produced by nuclear reactions

SUBTERRANEAN: under the surface of the earth

SUCCESSION: the natural ecological process which leads from simple groupings of plants toward complex botanical communities

TALLOW: hard fat obtained from animals, used in soap, candles

TERMINAL SPECIES: animals at the upper end of food chains ($q.v.$), usually large predators or omnivores

TERRESTRIAL: of the earth's surface

UNIFORMITY: having the same form, not varying

UNIQUE: without equal, rare or uncommon

UNPRECEDENTED: not done previously

VARIABILITY: the ability to change in form or behavior

Bibliography

Ardrey, Robert. *African Genesis: A Personal Investigation into the Animal Origins and Nature of Man.* New York: Atheneum, 1961.

Asimov, Isaac. *Building Blocks of the Universe.* New York: Abelard-Schuman, 1972.

———. *The Genetic Code.* New York: Signet Books, 1962.

Barbour, Ian G., ed. *Western Man and Environmental Ethics.* Reading, Mass., and Menlo Park, Calif.: Addison-Wesley, 1973.

Bateson, Gregory. *Steps to an Ecology of Mind.* New York: Chandler, 1972.

Bresler, Jack, ed. *Human Ecology.* Reading, Mass., and Menlo Park, Calif.: Addison-Wesley, 1966.

THE SPHERES OF LIFE

Calder, Nigel. *Eden Was No Garden, An Inquiry into the Environment of Man.* New York: Holt, Rinehart and Winston, 1967.

Carson, Gerald. *Men, Beasts and Gods: A History of Cruelty and Kindness to Animals.* New York: Charles Scribner's Sons, 1972.

Clapham, Wentworth B. *Natural Ecosystems.* New York: Macmillan, 1973.

Coon, Carleton. *The Hunting Peoples.* Boston: Little, Brown, 1971.

Darling, Frank Fraser. *Wilderness and Plenty.* New York: Ballantine Books, 1971.

———, ed. *Future Environments of North America.* Garden City, N.Y.: Natural History Press, 1966.

Dobzhansky, Theodosius. *Mankind Evolving.* New Haven, Conn.: Yale University Press, 1962.

Dubos, René. *Man Adapting.* New Haven, Conn.: Yale University Press, 1965.

———. *So Human an Animal.* New York: Charles Scribner's Sons, 1968.

Ehrlich, Paul R. *The Population Bomb.* Rev. ed. New York: Ballantine Books, 1971.

Ehrlich, Paul R. and Anne H. *Population, Resources and Environment.* San Francisco: W.H. Freeman and Co., 1972.

Eiseley, Loren. *The Immense Journey.* New York: Random House, 1946.

Elton, Charles S. *The Ecology of Invasions by Animals and Plants.* New York: Wiley, 1963.

Bibliography

Galston, Arthur W. *The Life of the Green Plant.* 2nd ed. Englewood Cliffs, N.J.: Prentice-Hall, 1964.

Giddings, J. Calvin. *Chemistry, Man and Environmental Change.* New York: Harper and Row, 1973.

Glass, David C. *Environmental Influences.* New York: The Rockefeller University Press, 1968.

Hafez, E.S.E., ed. *Behavior of Domestic Animals.* Baltimore: Williams and Wilkins, 1962.

Hardin, Garrett. *Population, Evolution, and Birth Control.* 2nd ed. San Francisco: W.H. Freeman, 1969.

———. *Exploring New Ethics for Survival: The Voyage of the Spaceship Beagle.* New York: Viking, 1972.

Huxley, Julian. *Evolution in Action.* New York: Harper and Row, 1953.

———, ed. *Evolution as a Process.* London: George Allen & Unwin, 1954.

Kormondy, Edward J. *Concepts of Ecology.* New York: Prentice-Hall, 1969.

Leopold, Aldo. *A Sand County Almanac.* New York: Oxford University Press, 1949.

Lorenz, Konrad. *On Aggression.* New York: Harcourt, Brace & World, 1966.

———. *Studies in Animal and Human Behavior.* 2 vols. Cambridge, Mass.: Harvard University Press, 1970, 1971.

Mayr, Ernst. *Animal Species and Evolution.* Cambridge, Mass.: Harvard University Press, 1963.

Morris, Desmond. *The Human Zoo.* New York: McGraw-Hill, 1969.

———, ed. *Primate Ecology.* Chicago: Aldine, 1969.

Nash, Roderick. *Wilderness and the American Mind.* New Haven, Conn.: Yale University Press, 1967.

Odum, Eugene P. *Fundamentals of Ecology.* 3rd ed. Philadelphia: Saunders, 1971.

Ortega y Gasset, José. *Meditations on Hunting.* Trans. Howard Wescott. New York: Charles Scribner's Sons, 1972.

Portmann, Adolph. *Animal Forms and Patterns.* New York: Schocken Books, 1967.

Potter, Van Rensselaer. *Bioethics: Bridge to the Future.* Englewood Cliffs, N.J.: Prentice-Hall, 1971.

Romer, Alfred S. *The Vertebrate Story.* Chicago: University of Chicago Press, 1971.

Ross, Herbert H. *Understanding Evolution.* Englewood Cliffs, N.J.: Prentice-Hall, 1966.

Russell, Claire, and Russell, W.M.S. *Human Behavior, A New Approach to Human Ethology.* Boston: Little, Brown, 1961.

Scully, Vincent. *The Earth, the Temple and the Gods.* New Haven, Conn.: Yale University Press, 1962.

Sewell, Elizabeth. *The Human Metaphor.* South

Bibliography

Bend, Ind.: University of Notre Dame Press, 1964.

Shepard, Paul. *Man in the Landscape.* New York: Knopf, 1967.

———. *The Tender Carnivore and the Sacred Game.* New York: Charles Scribner's Sons, 1973.

Shepard, Paul, and McKinley, Daniel, eds. *The Subversive Science.* Boston: Houghton Mifflin, 1969.

Sinnott, Edmund W. *The Biology of the Spirit.* New York: Viking, 1955.

Thomas, William L., ed. *Man's Role in Changing the Face of the Earth.* Chicago: University of Chicago Press, 1956.

Thompson, D'Arcy. *On Growth and Form.* Abridged ed. London: Cambridge University Press, 1961.

Thorpe, W.H. *Learning and Instinct in Animals.* Cambridge, Mass.: Harvard University Press, 1963.

Tiger, Lionel, and Fox, Robin. *The Imperial Animal.* New York: Holt, Rinehart and Winston, 1971.

Tinbergen, Niko. *The Study of Instinct.* New York: Oxford University Press, 1951.

Ucko, P.J., and Dimbleby, G.W. *The Domestication and Exploitation of Plants and Animals.* Chicago: Aldine, 1969.

Wagner, Richard H. *Environment and Man.* New York: Norton, 1971.

Wheeler, Reuben. *Man, Nature and Art.* New York: Pergamon, 1968.

White, Lynn, Jr. "The Historical Roots of Our Ecological Crisis," *Science* 155:1203–1207 (1967).

———. *Machina Ex Deo.* Cambridge, Mass.: M.I.T. Press, 1968.

Wynne-Edwards, V.C. *Animal Dispersion in Relation to Social Behavior.* New York: Hafner, 1962.

Index

abortion, 74
abundance, and social disruption, 74
adaption to environment:
  of animal species, 47–48
  and evolutionary survival, 82, 93, 100–101
  human, 61; mental, 84–85
  and the right to life, 96
agriculture:
  and ecology, 42
  effects of, 19
  and noosphere, 57
  See also farming
Alaska, 19
animals:
  behavior of, 83–84
  carnivorous, 50–51
  climax, 48
  endangered species of, 91, 96
  extinctions of, 47
  herbivorous, 50
  migratory, 48
  omnivorous, 51
  pioneering, 47–48
  specialization of, 48
  value to humans, 54, 56
anthropology, 56
archaeology, 56
astronauts, 36
atmosphere, 31–37
  "greenhouse effect," 36
  sources of, 31

Index

behavior:
  abnormal, 72
  adaptations of, 82–83
  animal, 52–53, 83–84
  antisocial, 76
  ethical, 95–96, 100
  ritual, 55
Bible, 37, 58
biosphere, 12
  plant life, 39–45
  animal life, 47–53
birth control, 77
Boulding, Kenneth, 63
brain, human, 84–85
  evolution of, 91
  metaphors of, 100
British Columbia, 18

carnivores, 50–51
cereal grains, 18
cesium, 21
chance, 80–82
Chardin, Pierre Teilhard de, 55
civilization, 12–13, 56–57, 84, 88
  role of fire in, 66
climax:
  animals, 48
  ecosystems, 41–42
Colorado, 19
communication, animal, 55
communities, botanical, 41–44

conservation of energy, 64, 67
corn blight, 88
cosmic radiation, 11–12
crisis, environmental, 7–8, 13–14, 60, 84
  and energy, 63
crust, earth's, 8–10, 19
culture:
  achievements of, 77
  noosphere, 12–13
  role of fire in, 66
  traditions of, 60–61, 79, 80, 83, 98–99
customs, *see* laws and customs

dams, effects of, 20
Darwin, Charles, 60, 79
Desert:
  North American, 19
  origins of, 18
  Sinai, 18
despotism, 93
dignity, 60
disease:
  and overcrowding, 76
  and population fluctuations, 72
  and uniformity, 88
diversity, 87–94
  and adaptability, 88–90
  bacterial, 44
  ecological, 42–44, 92
  evolutionary, 80

Index

and freedom, 93
genetic, 91–92
and love, 93–94
of species, 91
tolerance for, 102
domestication:
  animal, 53, 56
  human, 101
  plant, 42, 56
dominion over nature, 57

earthquakes, man-caused, 19–20
ecology:
  attitudes toward, 48
  defined, 12
  diversity of, 42
  and economics, 63, 67
  and farming, 42
  as metaphor, 100
  plant, 42
  as thought process, 61
economics, 63–64, 66
  and ecology, 67, 70
  of ecosystems, 64
  of solar energy, 67–68
ecosphere:
  defined, 13–14
  human influences upon, 58–60
ego, 60
elm disease, 75
endangered species, 91, 96
energy:
  conservation of, 64, 67
  conversion of, 64
  crisis, 63, 69
  electrical, 67
  in food chains, 50–52
  geothermal, 68
  limits of, 70
  paths of, 64
  solar, 64–67
  tidal, 68
epidemic diseases: 72, 76
  in plants, 88
erosion, 18–19
ethics:
  defined, 95–96
  environmental, 95–102
evolution, 60, 79–85
  and diversity of species, 91
  human, 80
extinction:
  animal, 47, 53
  and evolution, 82
  as human moral problem, 96

families, animal and human, 83–84
farming:
  and ecological simplification, 42
  as metaphor, 101
  and soil erosion, 16-18
  See also agriculture
fire, 66
fittest, survival, of, 82–83

Index

food chains, 12, 40
  energy in, 51, 64
  and humans, 56
  interdependency of, 52
fossil fuels, 40, 66, 68
fossil records, 80
freedom, 93

garden, metaphors of, 98–99
gas, natural, 20–21
golden rule, 96
grazing, 16–18
ground water, 28–29

Hanford, Washington, 21
heat, 64, 66–68
herbivores, 50
history:
  evolutionary, 85, 92
  human, 56, 58, 83
humility, 23, 61, 102
hunting, 53, 56
hybrids, 92
hydrologic cycle, 24
hydrosphere, 10–11, 23–29

icecaps, polar, 36
imagination, 55
industry:
  and energy use, 67–68
  and heat loss, 66
insecticides, 27, 75, 88–90

instincts, 52
inventories, ecological, 42–44

lakes, artificial, 20, 36
language, 55, 84
laws and customs, 57–58, 67–68
Lebanon, cedars of, 18
lichen, 50
lithosphere, 15–22
  defined, 8–10
  stability of, 21–22
London, 33
Los Angeles, 33
love, 94

machine, metaphor of, 99–100
manufacturing, 66
master race, 101–102
mathematics, 55
mating behavior, 72–74, 76
mental balance, 76
metaphor, 57, 98–100
mind, human, 55, 60, 62
  metaphors of, 98
  survival value of, 84
miscarriage, 74
mobility, animal, 48
moon, 36
mortality, infant, 76–77
myths:
  of creation, 57

120

Index

and ethics, 98–100
of water, 27

nature, 98–99, 101
  conquest of, 102
  human attitudes toward, 57–58
  indifference of, 82
  love of, 93
natural selection, 79–82, 96
neurosis, 76
niche, ecological, 51–52
nitrogen, 32
noosphere, 55–62
  defined, 12–13, 55
  and ecology, 56, 60
nutrition:
  animal, 50–51
  and crude oil, 68–69

oceans, 36
oil:
  as energy, 66
  extraction of, 20–21
  as food source, 68
  origins of, 40
omnivores, 51
opportunism, evolutionary, 82–83
overcrowding, effects of, 72–74, 76
oxygen:

sources of, 31–32;
transfer of, 33

petroleum, as food source, 68
photosynthesis:
  defined, 31
  effects of, 39–40
  and solar energy, 64, 67
  symbolism of, 45
phytoplankton, marine, 33
pioneering species:
  animal, 47–48
  plant, 40–41
plant life, 39–47
plants:
  pioneering, 40–42
  symbolism of, 45
plutonium, 21
pollution:
  adaptations to, 61–62
  air, 26, 33–36
  moral, 95
  and overpopulation, 75
  in plants, 50
  water, 24, 27, 28
populations:
  animal, 71–72
  and cultural achievements, 77
  explosions, 72
  human, 44, 54, 68–69, 75
  optimum, 76–77
  regulation of, 74, 76
precipitation, 24–26

Index

predators, 48, 72, 75, 99
pride, human, 58–61, 85
property, concepts of, 15, 57, 60, 98
psychology, 60, 72, 74

radiation, 11, 21
recycling, 69
religion, 60, 77, 96
resources:
    depletion of, 66, 69–70, 88
    as metaphor, 101
    nonrenewable, 66–68
respect for life, 102
respiration:
    components of, 32
    diseases of, 34, 69
    evolution of, 91
    in plants, 28
responsibility, ethical, 96, 98, 100
right to life, 54, 96
rights, natural, 57–58

San Francisco earthquake (1906), 20
shales, oil, 21
simplification of ecosystems, 87–88
sin, 98–99
Sinai, wilderness of, 18
sky, symbolism of, 37
slavery, human and animal, 101–102
smog, 61

snow, 26
social structure, animal, 53
soil:
    diversity of, 44
    erosion, 16–19
    formation of, 26, 41
    in lithosphere, 8–10
Spaceship Earth, 99
specialization, ecological, 48, 51
species:
    endangered, 91, 96
    evolution of, 80–85
    numbers of, 91
stability:
    of ecosystems, 48
    of land forms, 21–22
    of populations, 72
status, 58–59
strontium, 21
succession, ecological, 40–42, 47–48
sun, 64
    see also photosynthesis
survival, 82–83, 96
systems:
    ethical, 95
    of life, 8–14
    stable, 92
    symbolic, 55

technology, 60–61, 67–68, 70, 77, 84, 88–90
terminal species, 52
time:

Index

cyclic, 80
evolutionary, 79–80

uniformity:
 genetic, 88–90
 social, 93
uniqueness, human, 83

variability, genetic, 93
 see also diversity

wastefulness, 87
wastes, nuclear, 19–21
water, 10, 23–29
weather, 34–37
weeds, 41, 92
wilderness, 93

About the Author

Joseph W. Meeker is senior tutor, Humanities, at Athabasca University, Edmonton, Alberta, Canada, where he is engaged in developing a curriculum in which humanistic learning will be correlated with the sciences to develop an understanding of the relationships between human culture and the natural environment. He is also the author of *The Comedy of Survival: Studies in Literary Ecology.*